Louis Elsberg

The Throat and it's Functions

Louis Elsberg

The Throat and it's Functions

ISBN/EAN: 9783744678742

Printed in Europe, USA, Canada, Australia, Japan

Cover: Foto ©berggeist007 / pixelio.de

More available books at **www.hansebooks.com**

THROAT AND ITS FUNCTIONS. ˙ ɪ

A LECTURE.

LADIES AND GENTLEMEN :—Many of you know, but most of you probably do not know, that the Academy of Sciences, under whose auspices we are here assembled, is one of the oldest scientific societies among us. Under the name of " Lyceum of Natural History," it has existed since the early part of this century, and for over fifty years has been publishing from time to time its annals, " to record" as was said in an old advertisement, " new and valuable facts in natural history, and to advance the public good by the diffusion of useful knowledge."

About three years ago, the Lyceum, like a young lady on getting married, changed its name, to increase its sphere of action and usefulness. Last year, the Academy, recognizing that a knowledge of the advances of science should not be limited to the narrow circle of its own votaries, established a course of popular scientific lectures, three of which were delivered in this hall. The object of this course was announced to be : " To awaken and diffuse an interest in science among the community at large, and to make the work of the Academy more widely known, and its claim to public sympathy and support more highly appreciated."

In furtherance of the same object, the Academy has now been enabled to arrange for six lectures, the first of which was

delivered a week ago, and to the second of which I ∎∎∎ te you∎
∎nt attention this evening.

∎ I have professionally the repute of being a ∎
of cut-throat, I have chosen for my subject, the T∎

If you look in *Webster's Dictionary* for a defini∎∎∎∎ ∎∎ ∎∎∎
word "Throat," you will find that in ship-builder's and mar-
iner's language, it is "the inside of the knee-timber at the middle
or turns of the arm," and "that end of a gaff which is next the
mast;" also "the rounded angular point where the arm of an
anchor is joined to the shank." In chimney-sweep's language
it is "the part of a chimney between the gathering, or portion
of the funnel which contracts in ascending, and the flue." But
I am not going to talk to you in either of these languages. I
mean the throat which is an integral part of our own body.

Usually, in speaking of a person's head, we include both the
posterior portion which is the head proper, and the anterior
portion or face. So in speaking of the neck (*i.e.*, the part of the
body between the head and trunk), we include the posterior
portion or neck proper, and the anterior portion or throat. The
throat holds the same relation to the neck, that the face does
to the head, and as there is no distinct boundary line between
the face and head, so there is none between the throat and
neck. Furthermore, just as the face contains in the mouth
and nose the commencements of the passages for food and
breath, so the throat contains their continuations. It is this
fact which gives the throat its prominence and importance in
the animal economy.

Swallowing and breathing are the two most momentous functions for life, and the actions of the throat in performing these

and that other function, of which the throat is the special organ, namely, the production of *voice*, together with various modifications of these functions, constitute our theme. What I have to say, falls therefore, under the three heads, swallowing, breathing, and phonation.

I. SWALLOWING.

During health, swallowing is performed so quickly and so easily, that it seems a very simple process ; you bring a morsel of appropriate food or a drink under your nose, your jaws move apart, your mouth receives it, a little muscular action follows, —and it has gone on its way to the stomach. Perhaps you experience a good deal of pleasure in the process, if your appetite be good and the morsel delicious ; but you would hardly believe that the act of swallowing is exceedingly complicated, that its investigation has engaged the patient research of a large number of physiologists for more than 100 years, and that the mystery of its mechanism has been solved, and the process thoroughly understood only within the last ten or fifteen years.

The parts of the body specially engaged in swallowing are : the back of the tongue, the soft palate, the fauces, the larynx, the pharynx and the œsophagus or gullet.

I should be very sorry to inflict tedious anatomical details upon you, but I must make you a little acquainted with these parts.

If you look either into your own mouth widely opened befo e a mirror, or into a friend's, you will see on the floor of the moutu, the tongue, and on the roof beyond the hard palate a curtain or veil, hanging obliquely downward and backward, with

a small tongue-like appendage in the middle. This veil
is the soft palate, and the little appendage in the middle
is the uvula. From each side of the veil, two folds of
the mucous membrane emerge, one going forward to the side
of the tongue and one backward. These are the arches of the
palate ; and the space included between these palatine arches is
called the fauces. Beyond the fauces, the space at the back
of the throat is called the pharynx.

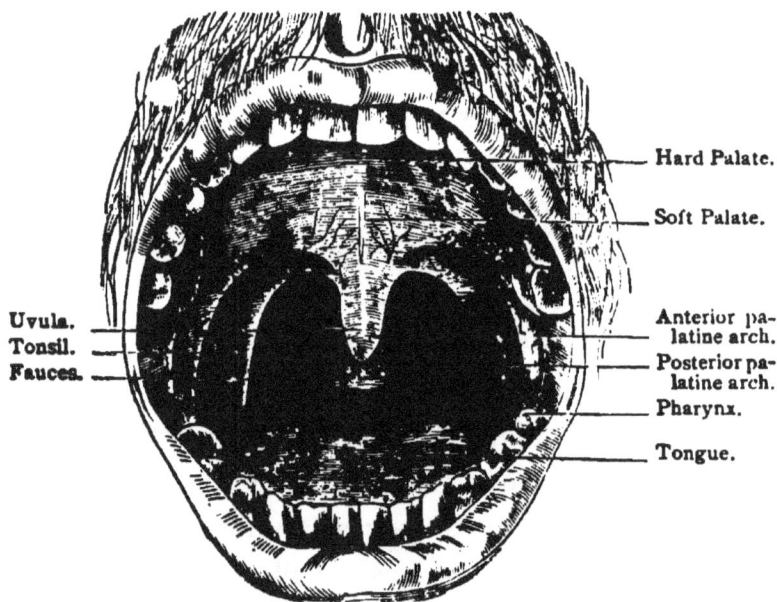

FIG. I. VIEW OF PARTS SEEN WHEN THE MOUTH IS WIDELY OPENED.

The tonsils, of which all of you have doubtless heard, are
the two almond-shaped bodies, one situated on each side of
the fauces in the triangular recess between the anterior and
posterior arch of the palate.

Fig. i is a representation of what is seen when the jaws are widely separated and the tongue is lying flat on the floor of the mouth. The roof of the mouth is the **palate** comprising two portions, viz.: the **hard** and the **soft palate** (*velum*) ; the tapering appendage at the middle of the latter is the **uvula**. On each side of the base of the uvula, the soft palate forms two crescentic folds, of which the front and narrower goes to the side of the tongue, and the other, the broader and longer one, to the side of the pharynx ; the first is called **palato-glossal** or **anterior palatine arch**, the second **palato-pharyngeal** or **posterior palatine arch**. The cavity of the mouth is held to terminate at an ideal plane passing through the anterior palatine arches ; beyond this, and to a plane passing through the posterior arches, the straits or passage leading to the back of the throat are called the **fauces** (*isthmus faucium*).

The **pharynx** is a cavity of very irregular shape extending from the base of the skull in front of the vertebral column or back-bone down to a level with the fifth vertebra of the neck. Below this level its continuation is called **œsophagus**. It is divisible into three portions, the middle of which is behind the mouth and fauces, and is called **pharynx**, in the restricted sense, or **oro-pharynx** to distinguish it from the upper portion or **naso-pharynx** (Fig. v, **A**) and lower portion or **laryngo-pharynx**. (Fig. v, **C**).

If you examine Fig. v, or the simple diagram Fig. ii, (say at the intersection of the dotted and unbroken lines) you will see that, aside from the track comprising from behind forward the pharynx, fauces and mouth, there are, at the back of

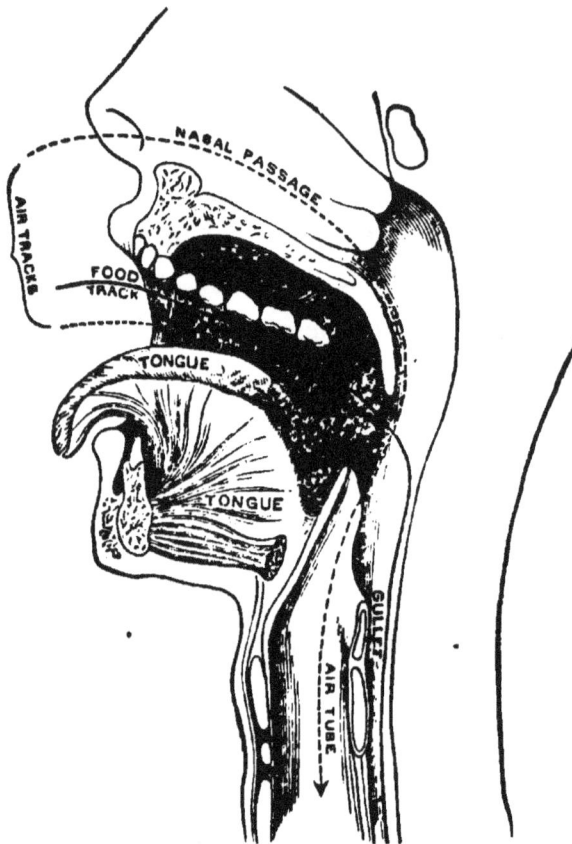

FIG. II. DIAGRAM SHOWING THE FOOD AND AIR TRACKS.

the throat, several passages leading in different directions, viz. : one upward and forward into the nose, another downward and in front into the windpipe, and a third downward and behind, into the gullet.

The last is the passage to be taken by the food ; and if you consider that this food-pipe is situated behind the wind-pipe, you will understand that some provision must exist to prevent the food from going the wrong way.   That provision is simple, yet beautiful, and of almost unfailing efficacy. The violent coughing and choking induced when food accidentally (or perhaps by our own negligence of the protection provided) passes into the wind-pipe, are but a specimen of the serious evils, the unpleasantnesses and the dangers to life, which would be continually occurring were it not for the following device. The top of the wind-pipe, which is called the larynx and of which I shall have to tell you a good deal more before the end of the lecture, is provided with a somewhat spoon-shaped or oval lid, (see Fig. iv, v or vii), called the cover-cartilage, or epiglottis (from two Greek words, *epi* upon and *glottis* the mouthpiece of a flute, because it shuts down *upon* what we shall learn is the *organ of the voice*). This lid has its hinge in front, and projects upward from behind the root of the tongue (see Fig. v, or the diagram Fig. ii. But it is more than a simple lid ; and it is not the only thing which prevents food from passing into the wind-pipe, for, in fact, in cases in which the whole upper portion of the epiglottis has been destroyed by disease, patients become able to swallow well enough without it. The sides of the upper portion of the larynx approach each other as though it was being compressed laterally, and the lower projecting portion of the epiglottis—its so-called cushion—closes over the entrance into the lower laryngeal cavity. There are therefore here *three safeguards*, and you can easily understand now how food can pass from the back of the tongue into the food-pipe without falling into the wind-pipe.

The passage of the food upward into the nostrils is prevented by the soft palate being raised and going backward, and the back wall of the pharynx coming forward, so that the two surfaces meet and shut off the nasal cavity ; besides which, the posterior palatine arches to some extent approximate, like side-curtains, and form inclined planes, along the under surface of which the morsel descends.

The act of swallowing is divisible into two periods, the first of which carries the aliment to the gullet, and the second to the stomach. The tongue by successive contraction and elevation from the tip to the base pushes the food toward the pharynx, and there are provided between the tongue and the epiglottis two pits or depressions, called valleculæ or little valleys, (see Fig. ii or v), to act as receptacles for overflow. This is the first place where foreign bodies are caught and detained.

At the same time that the nasal cavity is shut off, as just explained, the tongue-bone (see Figs. iv and v), and larynx are raised, carrying with them the floors of the valleculæ, so that these disappear, and the larynx is pushed, as it were, under the arched tongue ; and the bolus must pass over the top (which is really the front) surface of the epiglottis into the lower pharynx, where another pair of reservoirs for overflow is provided, called the pyriform or pear-shaped sinuses (see Fig. viii). This is another place where foreign bodies are easily caught and detained.

The bolus at this moment presses the larynx forward, but by the contraction of the constrictor muscles of the pharynx, is itself pushed into the œsophagus, and the pyriform sinuses are emptied by the action of the palato-pharyngeal muscles. You see, all this is the work of the muscular contraction of the tongue, soft palate, fauces, larynx and pharynx, and constitutes the first period of swallowing. The second is the period during which the parts hitherto contracted, relax. The larynx descends and thereby pushes the bolus a little lower, and once in the œsophagus the course of the food is easy enough. The œsophagus is simply a round tube made up of two rows of muscular fibres, the outer one longitudinal and the inner transverse and circular, with a soft moist lining membrane, which facilitates the transmission of the contents. When not expanded by the bolus it is closed; *i.e.*, its back and front walls are in contact; its lining membrane is then thrown into longitudinal folds, so that a cross section is somewhat star-shaped. Its position in relation to the back-bone is shown in the diagram Fig. iii.

The œsophagus reminds one of the good and noble people that are sometimes met with in the world, whose life is spent in doing good to others without a thought of self. It has been calculated that it carries more than two thousand pounds of food a year to the stomach, retaining nothing for itself.

The bolus, when introduced into the gullet, always assumes the shape of an egg with the pointed end upward. Even a liquid—for instance, a mouthful of water—gets this shape. The upper part of the tube is stimulated to contract by the presence of the mass, this contraction pushes the mass down-ward; the portion of the gullet now reached, contracts in its turn and propels the mass further; and so on in succession, till it arrives at the stomach. This successive or so-called peristaltic movement is easily observed in horses while they are drinking, as an undulatory or wavy contraction which proceeds rapidly along the tube ; and in the human subject, the sounds which this movement produces have recently supplied us with a new and important means of diagnosis in cases of disease of the gullet—the so-called auscultation of the œsophagus. The whole process of swallowing is peristaltic, and the wonderful accuracy with which the parts are successively brought into apposition is seen in the fact that we can swallow an exceedingly small bolus, as a very minute pill or grain of seed.

It is often supposed that the morsel passes along the gullet by its own weight ; but to correct this error we need but recollect that in the horse and cow, for example, the mouth is on a level with the ground when feeding, and that the morsel is consequently propelled upward into the stomach *against* its own gravity. It is well-known also, and often made a matter of public exhibition, that a man can swallow, even liquids, when standing on the crown of his head, with the natural position of the stomach reversed.

The *modifications* of swallowing need not detain us long. Dysphagia or painful swallowing is always a symptom of some diseased condition of the throat. Choking while eating is also symptomatic of throat disease, but with the parts perfectly healthy it is caused, when the person laughs or talks, or breathes, while in the act of swallowing,—sometimes from mere inattention or a sudden shock or mental impression. I have been at a dinner table at which a gentleman got a piece of meat lodged in one of the valleculæ, which so completely pressed down the epiglottis that breathing was stopped. The gentle-man became dark in the face, and in all human probability

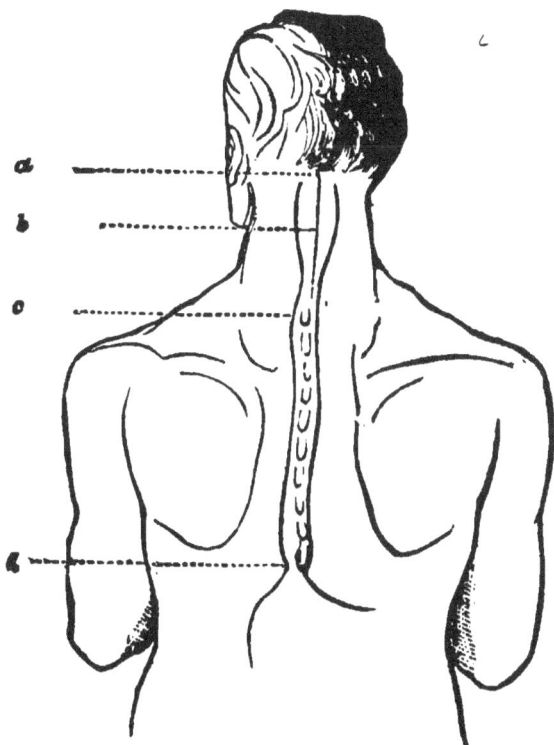

FIG. III. DIAGRAM OF THE PHARYNX, ŒSOPHAGUS, AND COMMENCE-
MENT OF THE STOMACH, SEEN FROM THE BACK.

a. Marks a level at about five-eighths of an inch below the
external protuberance of the occipital bone.

b. Shows the **pharynx,** which becomes narrow and forms
the **œsophagus** at about an inch above the lowest vertebra
of the neck.

c. Marks the position of the lowest or the so-called **prominent cervical vertebra.**

d. Shows the termination of the œsophagus and commence-
ment of the **stomach,** a little below the level of the lower
border of the **shoulder-blade,** *i.e.,* opposite the **ninth vertebra of the back.**

would have died within a few minutes, if the offending body had not been removed.

Gargling is a modification of swallowing a liquid, in which the epiglottis does not firmly close the laryngeal aperture and the expiratory current of air throws the liquid into vibrations, making the peculiar gurgling noise.

The mechanism of *eructation* of gas, *regurgitation* of liquids and solids, and *vomiting*, consists chiefly in an inspiratory depression of the diaphragm with holding the breath and contracting the abdominal muscles. This produces increased pressure in the cavity of the abdomen, and diminished pressure in that of the chest, acting on the gullet in the way of propulsion on the one hand and suction on the other. An anti-peristaltic muscular action of the œsophagus, which unquestionably occurs in ruminant animals, has not been proved to take place in man or carnivora. The windpipe and the posterior nasal opening are protected during eructation, regurgitation and vomiting, in the same manner as in swallowing. But the occlusion is usually less perfect : On this account the prickling effect of carbonic acid gas mixed with the air coming up f. i. after drinking sparkling liquids, may be distinctly felt in the nose ; and vomiting is followed as a rule by coughing, a reflex paroxysm to expel whatever may have found lodgment in the upper cavity of the larynx or near its aperture. The cough may, however, come also, alone, from irritation of those portions of the posterior laryngeal wall with which the· matter vomited necessarily comes into contact.

II.—BREATHING.

The action of the throat in relation to respiration consists in the conduction of the air ; in the regulation, in some degree, of its quantity and character, especially as to temperature ; and in the resistance afforded to the ingression of foreign bodies, which by their presence within the lungs or lower air tubes would be injurious to, or destructive of, life.

In quiet, natural breathing *the mouth is closed.* The air entering through the nose passes through the naso-pharynx into the middle portion of the pharynx, thence into the larynx, and so on through the windpipe and bronchial tubes into the lungs. It makes its exit through the same channels out through the nose. The cavity of the mouth is shut off from this passing current of air. The soft palate is in contact with the back of the tongue, and through the mutual approximation of the epiglottis and tongue, the valleculæ are closed and done away with, and in this position of the parts during quiet breathing no air can reach the mouth.

When we breathe wholly or partially through the mouth, the position of the parts involved is entirely changed. Breathing through the mouth takes place perfectly physiologically during violent, forced and deep respiration, that is, whenever the capacity of the nasal channels does not suffice for the volume of air to be inhaled ; also during talking and singing, and whenever the mouth is not closed and the uvula is drawn down. Abnormally, it occurs whenever, from catarrh or other disease, the nasal passage is obstructed or closed. The air, entering the oral cavity, passes through the fauces into the

pharynx and larynx and so on as before. The valleculæ may
then be more or less open, the soft palate may be raised and
the nasal cavity more or less shut off, which is especially the
case if immediately before, a vowel has been sounded and another is to be produced after the inspiration. In this case
the muscle of the uvula usually contracts.

If, from any cause, the uvula is relaxed, while it may touch
the wall of the throat behind or the raised back of the tongue
in front, the passing current of air may set it to vibrate sonor-
ously, *i.e.*, to making the inharmonic music called "snoring."

Do you want to know the conditions that produce, or favor
the production of, these sonorous or "snorous" vibrations ?

In the first place, the horizontal position during sleep, with
the mouth open ; and in the second place, very profound sleep,
as after severe bodily exertion, mental work or emotional
shock. The second on account of the greater degree of un-
consciousness, and therefore muscular relaxation ; and the
first because, in breathing through the nose, the air column
touches the posterior surface of the uvula alone, and this only
with little force ; but in breathing through the mouth, *both*
surfaces are energetically touched. All things that interfere
with natural and easy breathing during the sleep : therefore,
clothing tight around the throat or chest, heavy covering and
late suppers, especially exciting drink with well filled stomach
in corpulent persons, predispose to snoring. Whoever has
catarrh, or any disease interfering with nasal breathing, must
snore ; but very often after the necessity for breathing through
the mouth has passed away, habit keeps it up, and it is on this
account that a little invention of mine for the cure of snoring

was successful, and became very popular a dozen or more years ago. It consisted simply of a night-cap or muzzle, which forced the patient to keep the mouth shut. I do not mean to say, however, that it is impossible to snore with the mouth closed. It is only more difficult of accomplishment, but I have known inveterate adepts who have succeeded in spite of this and other obstacles. Sometimes I have prevented snoring by pulling the tongue out, sometimes by drawing the uvula forward and fastening it. If the last manœuvre is energetically carried out, snoring is made impossible.

I repeat that the natural mode of quiet breathing is through the nose ; mouth-breathing is an acquirement. A new-born infant would choke to death if you closed its nose ; it does not immediately know how to get air into the lungs through the mouth until after, by depressing the tongue, you have once made a passage for it.

George Catlin, the celebrated traveller among American Indians, became so thoroughly convinced that the difference between the healthy condition and physical perfection of these people in their primitive state, especially their sound teeth and good lungs, and the deplorable mortality, the numerous diseases and deformities, in *civilized* communities, is mainly due to the habit, common among the latter, of breathing through the mouth, especially during sleep,—that he wrote a book, entitled, " Malrespiration and its effects upon the enjoyments and life of man." In this book he says, " If I were to endeavor to bequeath to posterity the most important motto which human language can convey, it should be in the three words, 'shut your mouth.' In the social transactions of life this might have its beneficial results as the most friendly cautionary advice, or be received as the grossest of insults ; but where I would print and engrave it, in every nursery and on every bedpost in the universe, its meaning could not be mistaken, and if obeyed, its importance would soon be realized."

He also says, "It is one of the misfortunes of civilization that it has too many amusing and exciting things for the mouth to say, and too many delicious things for it to taste, to allow of its being closed during the day. The mouth therefore has too little reserve for the protection of its natural purity of expression, and too much exposure for the protection of its garniture ; but, *do keep your mouth shut* when you *read*, when you *write*, when you *listen*, when you *are in pain*, when you are *walking*, when you are *running*, when you are *riding*, and *by all means* when you are *angry!* There is *no person* but who will find and acknowledge *improvement* in *health* and *enjoyment* from even a temporary attention to this advice."

Again he says, "there is a proverb, as old and unchangeable as their hills, amongst North American Indians, 'my son, if thou wouldst be wise, open first thy eyes ; thy ears next, and last of all thy mouth, that thy words may be words of wisdom and give no advantage to thine adversary.' This might be adopted with good effect in *civilized* life ; he who would *strictly adhere* to it would be sure to reap its benefits in his *waking* hours, and would *soon find* the habit running into his hours of *rest*, into which he would *calmly* enter ; dismissing the nervous anxieties of the day, as he firmly closed his teeth and his lips, only to be opened *after* his eyes and his ears in the morning, the rest of *such* sleep would bear him daily and hourly proof of its value."

Catlin regards the habit of sleeping with the mouth open the most pernicious of *all bad habits.* The horrors of night-mare and snoring are according to him but the *least* of its evil effects. He thinks "for the greater portion of the thousands and tens of thousands of persons suffering with weakness of lungs, with bronchitis, asthma, indigestion and other affections of the digestive and respiratory organs," the correction of this habit is a *panacea* for their ills !

He insists that " *mothers* should be looked to as the first and principal *correctors* of this most destructive of human habits ; * * * and the united and simultaneous efforts of the civilized world should be exerted in the overthrow of a monster so destructive to the good looks and life of man. Every physician should advise his patients, and every boarding-school in existence and every hospital should have its surgeon or matron, and every regiment its officer, to make their nightly and hourly 'rounds,' to force a *stop* to so unnatural, disgusting and dangerous a habit ! Under the working of such a system, mothers guarding and helping the l. lpless, school-masters their scholars, hospital surgeons their patients, generals their soldiers, and the rest of the world protecting themselves, a few years would show the glorious results in the bills of mortality and the next generation would be a *regeneration* of the human race."

So much for the conduction of the air ; as to the regulation of its quantity and the protection to the lower respiratory organs afforded by the throat, I need here only refer to the valvular action of the palate, base of tongue, pharynx and epiglottis already described ; to the double valve arrangement in the interior of the larynx to be described ; to the hair in the interior of the nose ; to the possession, by the mucous membrane, of secreting glands to moisten it, and of cilia, or hairlike projections, moving in the direction of the outlet, to remove excess of secretion as well as foreign matter, and perhaps also to promote the mixing of the stagnant air in the lung with the freshly inspired air ; and finally, to the reflex production of cough.

Of the different *modifications* of breathing, so far as the throat is concerned, I may say that most of them are accompanied by the production of some sort of *noise*, as for instance, hawking and coughing, as well as the convulsive actions of laughing, weeping and sobbing ; each of these has a biography of its own, into the detailed consideration of which I cannot now enter. But I shall make a few remarks concerning some of them when explaining the production of vocal sound.

Sucking is an inspiratory act to produce a vacuum to draw into the mouth usually liquids, though sometimes also semisolids and æriform substances. *Sipping* is drawing in a liquid in small quantity by suction. In *smoking* of tobacco and other substances the smoke is drawn in by suction : The smoke is exhaled either through the mouth, by opening it suddenly, which is called *puffing*, or through the nose, but in either case without entering the larynx ; for at the very moment it approaches the epiglottis, expiration commences, except in the *Turkish mode of smoking*, which alone constitutes *inhalation* of the smoke. Some persons, however, *swallow* the smoke after sucking it into the mouth.

In *straining* and in every violent muscular effort, there is a closure of the parts, which confines the air within the lungs and throat. This is described by Shakespeare when he makes Henry V encourage his soldiers at the siege of Harfleur. (In King Henry V, act 3d, scene 1st.)

> " Once more unto the breach, dear friends, once more ; * * *
> In peace, there's nothing so becomes a man,
> As modest stillness, and humility ;
> But, when the blast of war blows in our ears,
> Then, imitate the action of the tiger : " * * * *

> " Now * * *
> Hold hard the breath, and bend up every spirit
> To his full height ! "

III. PHONATION.

All animals, except the lowest, possess some means of making themselves heard. Many have the power of emitting vocal sounds in which the practiced ear can recognize variations of cadence, character and power, conveying to other individuals of the same genus or species intimations of danger, of pleasure, of suffering or distress, of sources of nourishment, of desire, of affection, and perhaps many other emotions and suggestions, in a language which, in all except its most prominent and manifest meanings, is unknown to us. *Human voice and speech* constitute the grand distinguishing characteristics of humanity. They are the chief means of developing the mental faculties, and of communing with our fellowmen. They can embody all human perceptions and desires ; all that the mind can conceive, and all that the soul can feel. They can welcome, warn or persuade, attract or repel, excite or soothe, mourn or mock, sympathize or threaten, flatter, beg or command ; and in song, united with music, like Iris, the messenger of gods, they can carry to the souls of mortals unspeakable love, or passion, or peace.

Their mechanism somewhat resembles the musical instrument —the reed organ. In the organ there is a bellows to supply a current of air, a wind-chest or portevent to conduct it, a reed to vibrate, and a body-tube or resonance-pipe to augment and modify the sound produced. In the human voice- and speech-organ, the lungs are the bellows ; the bronchial tubes, from the smallest ramifications upward to the wind-pipe, are the wind-chest ; the larynx contains the reed ; and the space above the reed, including the upper cavity of the larynx and the cavities and adnexa of the pharynx, mouth and nose, constitutes the resonance-tube.

With some parts of this complex organ you are already acquainted.   I shall now describe in detail the hyoid bone, thyroid body, larynx, wind-pipe, with its ramifications, and lungs.

In Fig. iv, 1 is the **epiglottis** which belongs to the larynx, but projects above and behind the **hyoid** or **tongue bone** 2. On each side of the central portion or body of the latter is seen the **greater horn of the hyoid bone,** and at the junction of the greater horns and the body are the **lesser horns.** 1, 4 5 and 6 show the **larynx,** of which in a front view are seen three cartilages, viz.: epiglottis, thyroid and cricoid only ; the **thyroid cartilage** 4 shows a deep **notch** in the middle of its upper edge, into which the epiglottis 1 is inserted ; 5, 5 are the **superior horns** of the thyroid cartilage, and 3, 3 the

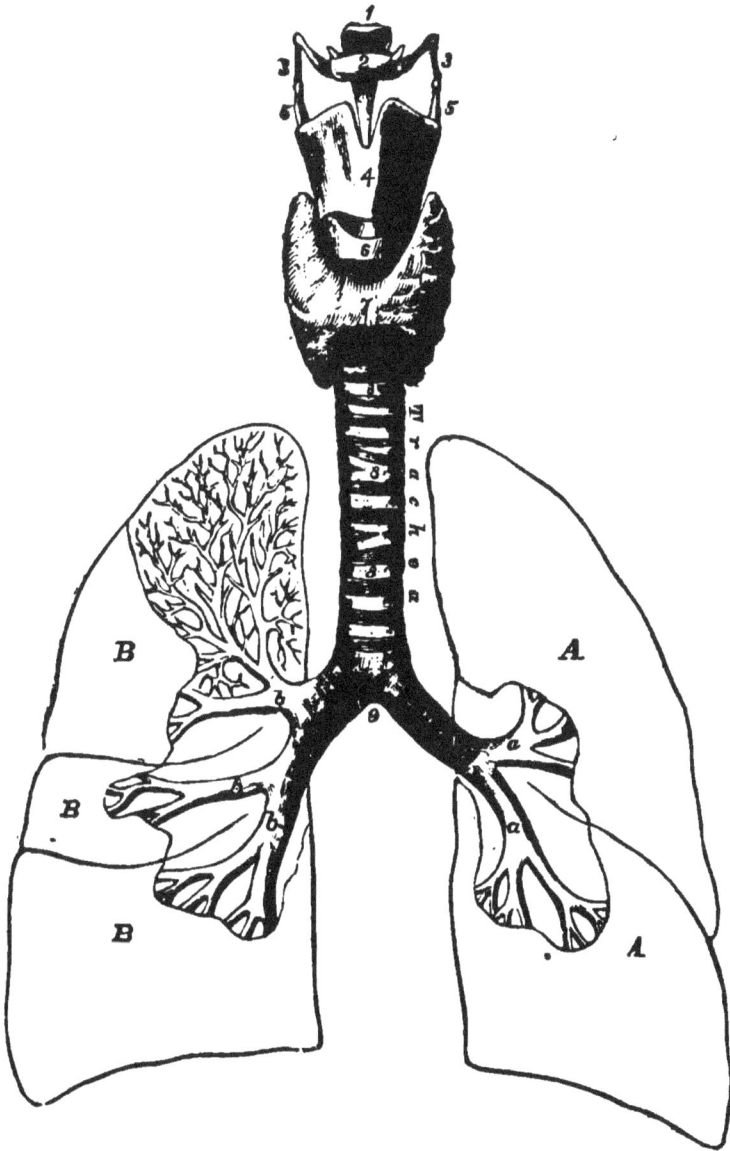

FIG. IV. A CONNECTED VIEW OF THE HYOID BONE, THYROID BODY LARYNX, WIND-PIPE WITH ITS RAMIFICATIONS, AND LUNGS.

thyro-hyoid ligament which fasten the ends of these horns
to the extremities of the greater horns of the hyoid bone, 6 is
the **cricoid cartilage**, which is seen to be narrow in front,
but with its upper edge rising obliquely : at its sides it is attached to the **inferior horns of the thyroid cartilage.**
Between the lower edge of the thyroid and the upper of the
cricoid cartilage a portion of the **elastic membrane** of the
larynx—the so-called thyro-cricoid membrane—is seen. Below
the cricoid cartilage is the **crico-tracheal ligament;** this
unites the larynx to the wind-pipe or **trachea,** which extends
down to the **bifurcation** 9.   At 8, 8, 8 are seen tracheal **cartilaginous rings.** 7 shows the **thyroid body,** its isthmus
being in the middle and its **lateral lobes** extending lower
down the trachea on the one hand, and on the other so high
up as to hide almost completely the **thyro-cricoid joint.**
At the bifurcation of the trachea 9, commence the **left
bronchus** 10 and the **right bronchus** 11. *A, A* show in
outline the two **lobes** of the **left lung** into which the **bronchial tubes** *a, a* are seen to enter. The three **lobes** of the
right lung are indicated by *B, B, B,* with the corresponding
bronchial tubes *b, b, b.* In the upper lobe of the right lung
is indicated in outline the manner in which the bronchial tubes
subdivide into smaller and smaller tubes, which finally terminate in **air-passages** and **air-cells** of the **primary lobules**
of the lungs, (see Fig. vi).

In Fig. v., the letter *A* shows the **naso-pharynx,** *B* the
oro-pharynx, and *C* the **laryngo-pharynx;** i to vii are
the **first** to **seventh cervical vertebræ;** 1 is the **frontal
sinus,** 2 the **sphenoidal sinus,** and 3 **occipital bone;** 4

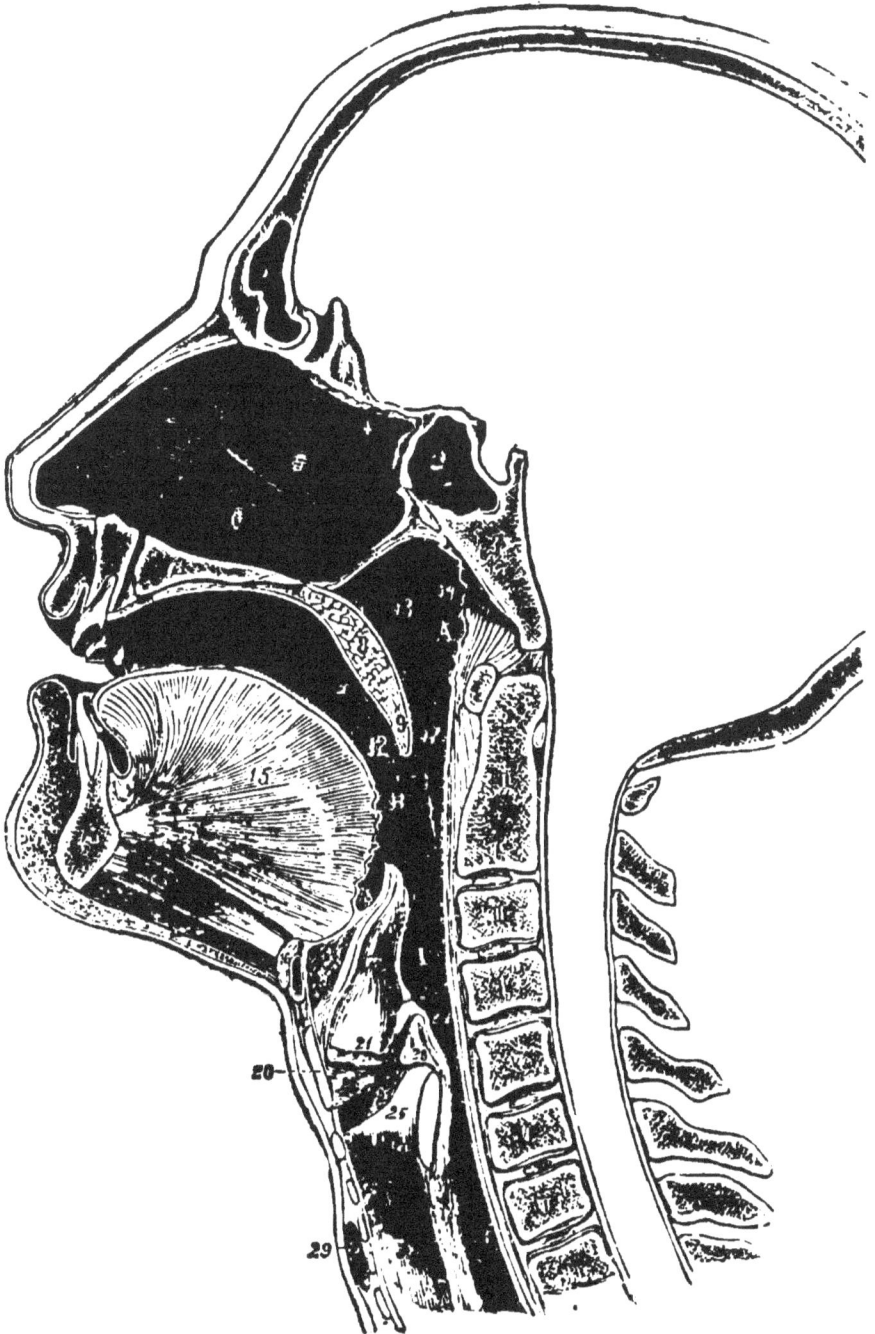

FIG. V. REPRESENTATION OF SECTION THROUGH HEAD AND NECK.

the superior and 5 the inferior turbinated process of the ethmoid bone; 6 the turbinated bone, 7 the hard and 8 the soft palate; 9 the uvula, 10 anterior palatine arch, 11 lower jaw bone, 12 tonsil, 13 orifice of the eustachian tube, 14 Rosenmüller's fossa, 15 tongue, 16 hyoid bone, 17 posterior palatine arch, 18 vallecula, 19 epiglottis, 20 thyroid cartilage, 21 ventricular fold, 22 vocal band, 23 arytenoid cartilage, 24 cuneiform cartilage, 25 cricoid cartilage, 26 anterior muscle, 27 supraarytenoid cartilage, 28 lateral muscle, 29 thyroid body, w ind-pipe, 31 food-pipe.

FIG VI. DIAGRAM OF TWO PRIMARY LOBULES OF THE LUNGS, MAGNIFIED.

A bronchial tube is seen to go to each primary lobule, of which a pair is represented, 2, 2, connected by fibro-elastic tissue; 3, 3, shows intercellular air passages; 4, air cells; 5, branches of the pulmonary artery and vein.

The lungs are essentially composed of groups of little elastic bags, called air vesicles or air-cells, each about the one-tenth of a line in diameter, which form a primary lobule (see Fig. vi, 2). Each lobule when sufficiently magnified shows some external resemblance to a bunch of grapes; in its structure it

represents the entire lung.   It consists, as the diagram Fig. vi
shows, of a minute bronchial tube (1), opening into an air-
passage (3), which communicates with a multitude of air-cells
(4).   When the chest is expanded, these air vesicles expand on
account of their elasticity ; a partial vacuum being produced,
air from the outside rushes in through the routes already
described to you.   When the chest contracts, the air-cells
contract or rather are squeezed together, and the air thus
expelled is supplied through the bronchial tubes and wind-
pipe to the larynx (see Fig. iv).

The bronchial tubes (Fig. iv, outer edge of upper *B;* Fig.
iv, 1), starting at the air passages (Fig. vi, 3), of the primary
lobules with a calibre of about the one-fortieth of an inch in
diameter, coalesce to form larger and larger tubes, until, in the
right lung (see Fig. iv, *b, b, b*), three of them, and in the left
lung (see Fig. iv, *a, a*), two, unite to form respectively the right
bronchus, which is about the fourth of an inch in diameter and
about an inch in length, and the left bronchus, which is narrower than the right, but about twice its length.

The trachea or windpipe (weasand), is a cylindroid tube,
(see Fig. iv) cylindrical in front and on the sides and flat behind (see Fig vii), about four inches long, from three-fourths
to an inch in breadth, and one-half to three-quarters of an inch
in depth.   It is composed of a series of incomplete cartilaginous rings, varying irregularly from sixteen to twenty in
number, connected by a fibro-elastic membrane, and is lined
with a mucous membrane continuous above with the lining of
the larynx and below with that of the bronchial tubes.   The
cartilaginous rings are somewhat horse-shoe shaped, because

they are imperfect at their posterior third, which gives a section of the tube the tunnel-shaped appearance seen in Fig. vii.
The posterior interval of the cartilaginous rings is occupied by
a loose fibrous membrane and an internal layer of pale, unstri-
ated muscular fibres in a transverse direction.

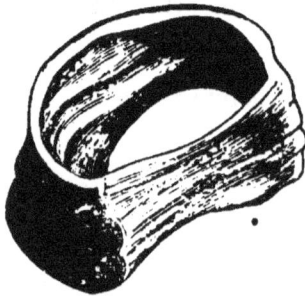

FIG. VII.   A PORTION OF THE TRACHEA.

The upper front portion, sometimes fully three-fourths of the
circumference of the trachea is grasped by the thyroid body.
(See Fig. iv, 7.)   This is a moderately soft, reddish organ,
which acts as a cushion and as a temporary blood reservoir.
It consists of a pair of lateral lobes united by a transverse
isthmus. (Fig. iv, 7.)   The isthmus usually covers the second,
third and fourth tracheal rings, but extends sometimes as high
up as the cricoid cartilage of the larynx or with a narrow projection even as high up as the hyoid bone, and exceptionally as
low down as the fifth or sixth ring.   The lateral lobes are
oblong oval, thicker below than above and usually of unequal
length ; they commonly measure about two inches, extending
from the sixth tracheal ring up to the lower part of the thyroid
cartilage.   The thyroid body is frequently larger in the female

sex than in the male and more liable to enlargement ; its permanent enlargement constitutes the affection called goitre.

The top of the windpipe, the larynx, (see Fig. iv, 1, 4 and 6)
is hung up in the fore part of the neck attached to the hyoid
or tongue bone. (See Fig. iv 2.) The hyoid bone itself, is
not directly connected with the other bones of the body ; but
is held moveably in its position at the root of the tongue and

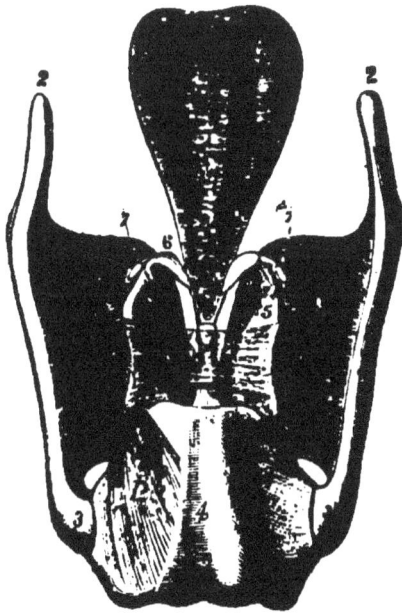

FIG. VIII. FRAMEWORK OF THE LARYNX, AS SEEN FROM BEHIND.

1, Epiglottis ; 2, 2, great horns, 3, 3, small horns of
the thyroid cartilage ; 4, plate of the cricoid cartilage ;
5. right arytenoid cartilage ; 6, left supra-arytenoid
cartilage ; 7, 7, sesamoid cartilages ; P, left posterior
laryngeal muscle ; T, transverse laryngeal muscle.

behind, and just under, the chin by numerous ligaments and muscles. In shape it resembles a horseshoe, and consists of a middle portion or body, two greater horns and two lesser horns. It supports the tongue and also the larynx, keeps the pharynx patulous and prevents the epiglottis from inclining too far forward.

The larynx is an irregularly shaped, but as to its two lateral halves symmetrical, open box, somewhat triangular above and narrow and cylindrical below. In size, adult larynges differ considerably, especially according to sex ; the extremes are that in the three dimensions : height, (not taking into account the length of either the upper horns of the thyroid cartilage, Fig. iv, 5, 5, or the epiglottis, Fig. iv, 1), width and depth, a large male larynx measures about an inch and nine or ten lines, and a small female larynx about an inch and one and a-half or two lines, the average male larynx measuring a few lines less, and the average female larynx a few lines more than these extremes.

The larynx consists of a framework of cartilages (pieces of gristle), which are held together by ligaments and moved by muscles, provided with blood-vessels, lymphatics and nerves, and lined first by a peculiar elastic membrane, and secondly by a mucous membrane continuous with that of the mouth, nose and pharynx above and trachea below. The cartilages composing the larynx are nearly always nine, and occasionally eleven, in number, three being single, and three or occasionally four, pairs. They are :

1. The basis or ring cartilage (cricoid) below. (See Fig. iv, 6 ; Fig. viii, 4.) This is narrow in front and rises up to be very

broad behind (Fig. v, 25). On each side it presents a surface for the joint of the shield cartilage. Its broadened posterior portion is called its plate.

2. The shield cartilage (thyroid), the form of which is nearly that of the cover of a book half opened, the back of the book representing the projection in front externally, known as "Adam's apple." On the posterior surface this is called the receding angle. A short projection of the shield cartilage called the lower horn, (Fig. viii, 3) is attached to each side of the basis cartilage forming a hinge joint; while the upper horns (Fig. iv, 5), are fastened by means of the thyro-hyoid ligament, (Fig. iv, 3), to the ends of the great horns of the hyoid bone. The sides of the shield cartilage flare out to make the pyriform sinuses. (Fig. viii, ; Fig. xi, 1, 1.)

3. The cover cartilage (epiglottis), which, being inserted into a deep notch of the shield cartilage in front, during swallowing descends, as you remember, like a lid or valve and closes the voice-box and windpipe, so that food passes over it into the food-pipe. (Fig. iv, 1 ; Fig. viii, 1, etc.)

4. The two pyramidal or moving cartilages (arytenoids), which are placed with a hollow base upon corresponding convexities on the top of the basis cartilage behind ; (Fig. viii, 5 ; Fig. xi, 3) making a ball and socket joint.

5. The two buffer cartilages (supra-arytenoids), one on the top of each arytenoid cartilage, to deaden or distribute pressure, preventing injury to the larynx, especially in swallowing hard morsels. (Fig. v, 27 ; Fig. viii, 6.)

6. The two prop cartilages (cuneiforms), which hold up the folds of the mucous membrane extending one on each side,

from the arytenoid cartilage to the epiglottis, constituting the
sides of the upper laryngeal aperture. (Fig. v, 24.) These
wedge-shaped cartilages help to keep this aperture patulous.
Occasionally these are missing in the human larynx, especially
when the aryepiglottic folds are tense without them, on account
of great development of the arytenoid cartilages.

The exceptional pair of cartilages (sesamoid cartilages), are
situated at the upper lateral edge of the arytenoids, (Fig.
viii, 7, 7) and when present facilitate somewhat the action of the
muscles that pull the epiglottis backward.

The anatomical name of each of these cartilages may be remembered by the following mnemonic device.   Think of writing the word " Coffee," and after having written a capital C as
the first letter, change your mind and write " tea " in capitals;
then add the small letter " c " for cream and " s " for sugar
and another " s " for an exceptional larger quantity of sugar
than that taken by most people.   (See Diagram, Fig. ix.)

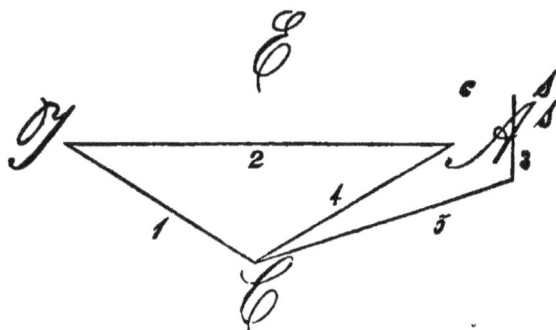

Fig. IX.   Diagram of Cartilages and Intrinsic Muscles of the
Larynx.

Then, *C* stands for **cricoid**, the basis cartilage; *T* for

thyroid, the shield cartilage ; *E* for epiglottis, the cover cartilage ; and *A* for arytenoid, the moving cartilages ; these are
the large cartilages, the small *c* stands for cuneiform, the
prop cartilages ; *s* for supra-arytenoid, the buffer cartilages ;
and the additional s for sesamoid, the exceptional cartilages.

By uniting the three lower letters by lines, drawing a transverse line behind the *A*, and uniting its terminus with the
lower letter *C*,—five of the· intrinsic muscles of the larynx,
which are the most important ones concerned in phonation,
may also be easily remembered. They are : 1, the anterior
(thyro-cricoid) muscles ; 2, the interior (thyro-arytenoid) muscles ; 3, the transverse (arytenoid) muscle ; 4, the lateral (lateral
crico-arytenoid) muscle ; and 5, the posterior (posterior crico-
arytenoid) muscles. All these except the arytenoid or transverse muscle are in pairs, *i.e.*, situated on each side of the
larynx ; and it is seen that all of them except the thyro-cricoid
or anterior are inserted into the arytenoid cartilages. Of the
functions of these various muscles it suffices here to say that
the posterior muscles by their contraction widen the space in
the interior of the larynx as I shall explain presently, while all
the others narrow it.

The interior of the larynx is somewhat hour-glass shaped.
(see Fig. x). It is divided into two cavities, an upper and a
lower, by two horizontal lateral projections which have a slit
or cleft-like space between them. These lateral projections
are the reeds of the vocal organ, called the vocal cords, or,
more properly speaking, vocal bands (see Fig. x, 9); and the
slit or opening between these projections is the vocal chink or
rima glottidis.

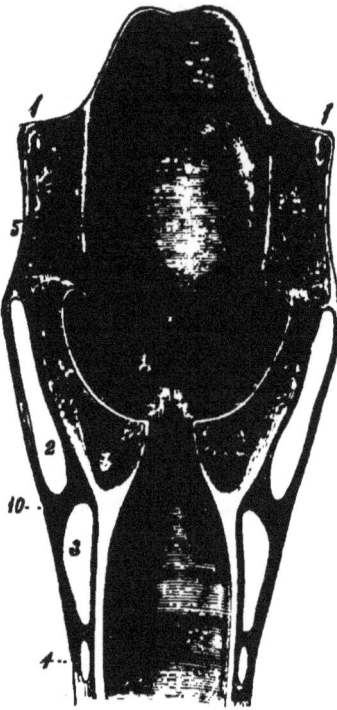

FIG. X. VIEW OF THE INTERIOR OF THE LARYNX, THE POSTERIOR
HALF BEING CUT AWAY.

I, 1 the greater horns of the **hyoid bone,** cut across ; **2
thyroid cartilages,** 3 **cricoid cartilages,** 4 **first ring of
the trachea** ; 5 the **thyro-hyoid membrane,** 6 **upper portion of the epiglottis,** 7 cushion of the epiglottis ; 8 **ven-
ticular fold,** showing above it the **wedge-shaped space** of
the **upper laryngeal cavity;** 9 left **vocal band,** *a, b, c*
showing the different portions of the **interior muscle ;** the
muscular tissue above belongs to the **muscles affecting the
position of the epiglottis.** The **wedge-shaped space
of the lower laryngeal cavity** is well shown.

This chink is correctly enough termed *rima glottidis*, literally chink of the glottis, but some call it simply glottis, although that word being derived from "glotta" meaning tongue, would be more appropriately applied, as GALEN did, to the projections themselves than to the opening between them. The action of the five muscles that I have described to you is exerted upon these lateral projections or vocal bands. The posterior muscles are abductors óf the vocal bands, *i.e.*, they draw them apart and thereby widen the chink between them ; the anterior, interior, transverse and lateral muscles are adductors of the vocal bands, *i.e.*, they bring about (together with other effects, such as variations of tension and of length and thickness), more or less approximation of the lateral projections, and thereby narrow and even obliterate the chink.

The hour-glass shape of the interior of the larynx is produced as follows : The upper surface of the projections is horizontal ; but from its free edge, each projection slopes obliquely downward, so that the space below is wedge-shaped, with the point upward. Above the projections, the lining of the cavity *i.e.*, both the elastic and the mucous membrane, is tucked in as it were, or folded upward, to form a hood or pouch in each side wall of the larynx. The pouch is conical in form and slightly curved from before backward, resembling a "phrygian cap ; " it is called the sacculus of the larynx. The entrance to the sacculus, an oblong fossa situated along nearly the whole length of and just above each lateral projection, is called the ventricle of the larynx ; and the fold of the membrane which is tucked in and forms the upper boundary of the ventricle is called the ventricular fold. The sacculus on each side being more or less filled with air, the fold which is its inner wall is, of course, pushed more or less inward and upward ; therefore the space above the projections is wedge-shaped as well as the space below, but with the point downward.   The double valve arrangement in the interior of the larynx to which I have already alluded as controlling the conduction of air is thus constituted.   The lower valve, formed by the more or less approximated lateral projections (vocal bands), obstructs or

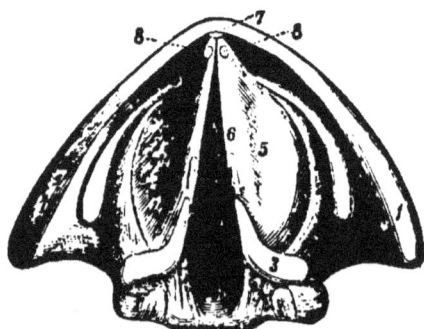

FIG. XI. VOCAL PORTION OF THE ELASTIC MEMBRANE OF THE LARYNX, IN CONNECTION WITH THE CARTILAGES, SEEN FROM ABOVE.

I. 1 Section of the thyroid cartilage ; 2 cricoid cartilage, 3 right arytenoid cartilage ; 4 crico-arytenoid ligament ; 5 elastic membrane of the larynx ; 6, 6 duplications of the elastic membrane, seen to extend from 7, the anterior vocal process to the two posterior vocal processes, which are the most anterior projections of the arytenoid cartilages.   These duplicatures or folds contain 8, 8 the two anterior vocal nodules, and 9, the two posterior vocal nodules.   Having on each side the interior muscle filling up the space to the lateral walls and being covered with mucous membrane, they constitute the vocal bands.

regulates the entrance of air, while the upper valve, formed by the ventricular folds, obstructs or regulates the exit of air.

About 140 years ago a French physician, Ferrein, gave the name "*chordæ vocales*" to the lateral projections in the interior of the larynx because he imagined their action to be analogous to that of tensioned strings. This name was unfortunately universally adopted and has survived the erroneous hypothesis on which it was founded. They are still called in English "*vocal cords*," in French, "*cordes vocales*," in German, "*Stimm-bänder.*" They certainly act like a tongue of a reed-pipe and not like musical strings or cords. I tried to reëstablish a more correct name, such as vocal reeds, vocal tongues, lips of the glottis, or simply glottes, names that had been used ages ago ; but finding it impossible to displace the Latin appelation mentioned, I have, in correspondence with the German word "Stimm-band," called them vocal bands, a designation that has found favor with several writers on the subject.

Each vocal band consists chiefly of its muscle, which is the interior or thyro-arytenoid muscle, and elastic tissue, which is a duplicature of the elastic membrane of the larynx, (see Fig. xi, 6) and a covering mucous membrane. It extends from the anterior vocal process, which is a little projection on the inside of the "Adam's apple," (Fig. xi, 7) to the posterior vocal process which is a little projection at the base of each moving or arytenoid cartilage. Near the free edge, its front portion usually contains, especially in the male larynx, a small dense nodule, which I call the anterior vocal nodule, (Fig. xi, 8), and its posterior portion, more often in the female than in the male sex, an elongated nodule which I call the posterior vocal nodule, (Fig. xi, 9). As the two vocal bands are attached in front in close contact and behind to two separate cartilages, the space between them must, when uninfluenced by muscular action, be of triangular form, (see Fig. xi). This position of the bands, which is the one found in the dead subject, is called the cadaveric position. When the bands are moved further apart, the base of the triangle is lengthened; when they are approximated, they become parallel and, then finally, the triangular space disappears. The manner of their adduction and abduction may be illustrated by supposing two fingers, say the middle and index, to represent them. The juncture of the fingers with the hand then represents the anterior angle of attachment of the vocal bands, the so-called anterior commissure, (Fig. xvi, 1). Supposing the fingers to be of the same length, the finger-nails would represent the place of posterior attachment, viz., the posterior vocal processes; let these be moved apart as far as possible and we have the position of greatest abduction something like that represented in **Fig. xvii** the more these are brought together or adducted, **the more** nearly parallel the fingers become and, of course, the more narrow becomes the space between them, representative of the *rima glottidis.*

You know musical cords or strings, as those of the guitar, violin, etc., are attached only at their two ends, so that they can freely vibrate between; the tongues or reeds of organs, accordeons, clarionettes and all other artificial reed instruments, are usually attached at one end only, so that they have three free edges; but the human reeds or vocal bands are attached on three sides and have only one free edge. Those of you who know what a large number of reed or organ pipes are needed in the organ made by man, to produce the notes of varying pitch and timbre, cannot fail to be struck with astonishment at the fact that in the organ in man's body a single reed-pipe, the larynx—by a wonderful power of variation inherent in itself—suffices for the production of the most various sounds. No musical instrument has ever been constructed by man that approaches in perfection or effectiveness that of the human voice.

Now, there is hardly any subject on which physiologists have differed more than they have, until recently, on the precise production of the voice. At the present day, however, it is perfectly understood and this is principally due to the introduction, about twenty years ago, of an instrument by which we can watch the whole process in living working order. This instrument is called the laryngoscope. The introduction of the laryngoscope we owe in the first place to the observations upon himself of the celebrated singing teacher Garcia, now living in London, and in the second place, to the independent investigations of two Austrian physicians, Turck and Czermak.

I doubt not, ladies and gentlemen, you will be astonished when I make you acquainted with the wonderful instrument, by the aid of which we can see and touch interior portions of the body that were heretofore impenetrably veiled from mortal gaze. This instrument has entirely revolutionized human knowledge of the throat in health and in disease. Its introduction is unquestionably "the most important improvement recently made in practical medicine." It has frequently enabled us to relieve suffering, to save life, to remove tumors

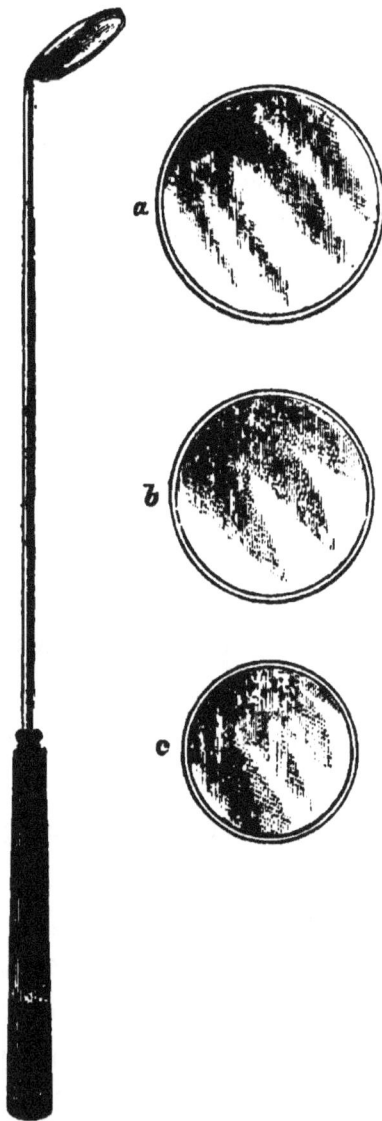

FIG. XII.—LARYNGEAL MIRROR, ONE-HALF OF ACTUAL DIMENSIONS; AND THREE MIRROR SURFACES OF ACTUAL SIZE.

which were the threatening yet unsuspected sources of sudden death, and to effect the miracle of making "the deaf to hear, and the dumb to speak." Yet the instrument and its employment are exceedingly simple. The essential apparatus is only a little looking-glass (see Fig. xii). In the words of Czermak, "a small flat mirror, having a long stem and being previously warmed, to prevent its being tarnished by the breath, is introduced into the mouth, widely open, as far as its back part. It is then held up in such a manner as to permit rays of light to fall upon it, to illuminate on the one hand the parts to be examined, and, on the other, to reflect the images of those parts into the eye of the observer." (see Fig. xiii) The warming of the little mirror is a necessary procedure, because the moisture of the expiratory breath otherwise condenses

upon it and dims its surface ; it is therefore dipped into water
of the proper temperature and then dried, or better still, is
warmed over a flame. In the latter case, as I pointed out
many years ago, a film first gathers upon the mirroring surface,
then clears away, and this is just when the mirror is ready for
use ; neither too cold to be dimmed nor too hot to be unpleasant to the throat.

The manner in which reflection in the mirror enables us to
inspect the interior of the larynx may be illustrated by placing

FIG. XIII.—MANNER OF HOLDING THE TONGUE AND THE LARYNGEAL
MIRROR.

the thumb behind the closely approximated fingers of the hand ; a beholder from the front cannot see the thumb-nail, although he can look in a horizontal direction beyond the hand ; but by means of the mirror held obliquely above it, the thumb-nail is readily brought into view.   In the same way, on looking directly into the throat, we can see only in a horizontal direction ; but by means of the little mirror we can look "round the corner," or in a vertical direction, down into the windpipe.   This is on the same principle as that of which some ladies avail themselves who have a mirror placed on the outside of the front window upstairs, to reflect the image of the street into the room, so that they can see who passes and who rings the bell, and be " at home " or " out," as they may please to instruct the servant, without danger of being themselves seen.

When we cannot sufficiently illuminate the mouth by the direct rays of the sun or other light, we must reflect light into the cavity.   This is done by a concave mirror, and thus our laryngoscope is complete.   I have combined these mirrors into a simple and efficient instrument for practical purposes, which is known in the profession as " Elsberg's pocket laryngoscope."

I have also had constructed a complete apparatus for artificial illumination, admitting of being attached to a gas—preferably argand burner—drop light, a so-called German student lamp, or any other convenient lamp (see Fig. xiv), which is a modification of Tobold's laryngoscopic arrangement.

Every singing teacher and elocutionist ought to practice the method of laryngoscopy.   To physicians, called upon to treat diseases of the throat, it is indispensible ; although I cannot deny that there are some who have failed to make themselves

FIG. XIV.—MANNER OF USING THE COMPLETE LARYNGOSCOPIC APPARATUS.

sufficiently familiar with it. In abnormal conditions, obstacles sometimes present themselves, making a thorough inspection difficult ; but ordinarily, in health, it is so easily accomplished that no one interested, professionally or as an amateur, in the cultivation or proper use of the voice for either singing or speaking, need to forego the important information and the indescribably great satisfaction which it grants. Any intelligent person, with an expenditure of less than an hour's time, and not much more than a dollar in money for the purchase

of a laryngoscopic mirror, may learn to inspect his own or
another's larynx.   The little mirror introduced into the mouth
as shown in Fig. xiii is the only essential apparatus needed to
examine some one else ; to look at one's self, one needs, of
course, in addition a larger mirror, to reflect the image of the
smaller one.   In Fig. xii are represented the actual sizes of the
mirrors I ordinarily employ ; the smallest usually for children.
The larger the mirror which the throat is taught to tolerate, the
more light it throws upon the part, and the more complete is
the image obtained.   Over-sensitivensss of the throat may be
overcome by touching frequently the various portions of it with
my "throat educator," made of wood or hard rubber, (Fig. xv).

FIG. XV.—THROAT EDUCATOR, ONE-HALF ACTUAL SIZE.

Suppose you are standing or sitting with your back to
a window, through which the sun-light enters, with a
looking-glass in front of you so arranged that you see the illu-
minated interior of your widely opened mouth.   If then, 1,)
you protrude and either with your will-power alone or with
this combined with the thumb and index finger of one hand
covered by a handkerchief (as is shown in Fig. xiii), hold the
tongue out and downward, while it is made to lie as flat as
possible on the floor of the mouth ; and 2,) you introduce with
the other hand the little mirror, previously (as I have already
described to you) warmed, and rest its back against the uvula,
at the same time that, breathing quietly and deeply, you utter

with each expiration a high-pitched sound "ai" (as in the word *fair*) ; you will, provided you depress or raise the handle of the mirror, or turn it a little more to one side or the other, as may be required, see the image of your larynx.

"If at first you don't succeed, try, try again." If you can

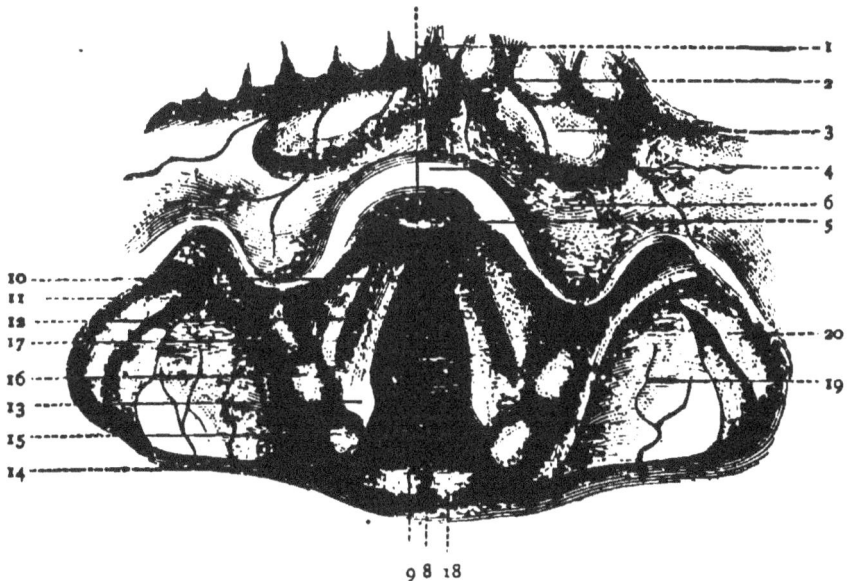

FIG. XVI. IMAGE OF THE LARYNX AND SURROUNDING PARTS, TWICE THE ACTUAL SIZE.

1, Base of the tongue ; 2, epiglottic frænum or middle glosso-epiglottic ligament ; 3, vallecula ; 4, epiglottis ; 5, cushion of the epiglottis ; 6, lateral glosso-epiglottic ligament ; 7, anterior and 8, posterior commissure of the larynx ; 9, rima glottidis ; 10, vocal band ; 11, ventricular fold ; 12, ventricle ; 13, posterior vocal process ; 14, arytenoid cartilage ; 15, supra-arytenoid cartilage ; 16, cuneiform cartilage ; 17, ary-epiglottic fold ; 18, posterior laryngeal wall, entrance to the œsophagus ; 19, pyriform sinus ; 20, hyoid fold of mucous membrane.

get some one who has already had some experience in laryngo-scopy (literally, the inspection of the larynx) to show you the image of his or of your own throat, you will learn the method very much more easily ; but if you must be your own teacher, you will certainly acquire it by carefully repeated examinations of your own and some one else's throat. To examine another by direct sunlight, you can dispense with the looking-glass and put him in the place of it (of course guarding his eyes from the sun's rays if they prove annoying). But it is pleasanter to examine with reflected light. To do this most simply, use my pocket laryngoscope which, being provided with a plane glass reflector, one concave one of short focal distance and another concave one of longer focal distance, held in the hand or attached either to the handle of the mirror or by means of a band or spectacle-frame to the forehead, is adapted to either sunshine, diffuse daylight or an artificial light. To use the complete laryngoscopic apparatus (see Fig. xiv) seat the person to be examined opposite to you in such a way that his right elbow can rest upon the edge of the table on which the lamp carrying the apparatus is placed. The centre of the illu-minating rays, your own eye and the patient's mouth must be on about a level. To examine your own throat with the same apparatus, take the place of the person to be ex-amined and place a looking-glass where your own eye was before.

Sometimes you can see through the open vocal chink into the windpipe ; sometimes as low down as the bronchi, as exhibited in Fig. xvii.

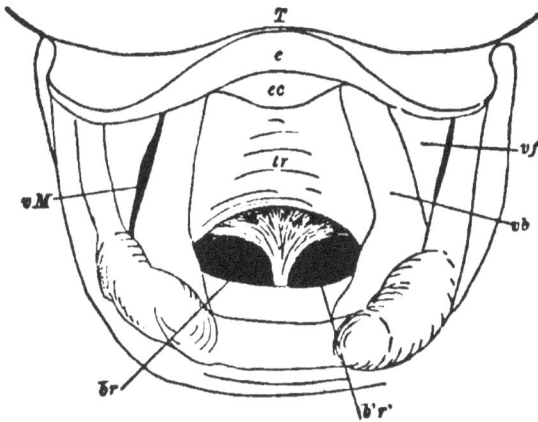

FIG. XVII. LARYNGOSCOPIC IMAGE SHOWING THE BIFURCATION OF THE TRACHEA.

T, tongue; *e*, epiglottis; *ec*, cushion of the epiglottis; *vb*, vocal band; *vf*, ventricular fold; *vM*, ventricle of the larynx; *tr*, trachea; *br* and *b'r'*, right and left bronchi.

The following three pictures represent cases of disease. The first (Fig. xviii) shows the vocal bands of a lady with a

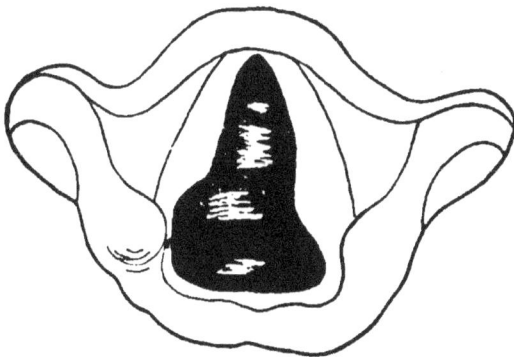

FIG. XVIII. LARYNX OF PATIENT WITH DOUBLE VOICE.

double voice, that is each sound of her ordinary conversational voice, and still more each *note* that she attempted to sing, was accompanied by a deeper, man-like sound, varying on the descending scale from being a second to a major third, or even a fourth or discordant sixth, of the higher one. The second picture (Fig. xix) shows the development in the vocal bands of a disease which is exceedingly common in this country in the eyelids, namely, trachoma or chronic granulations.

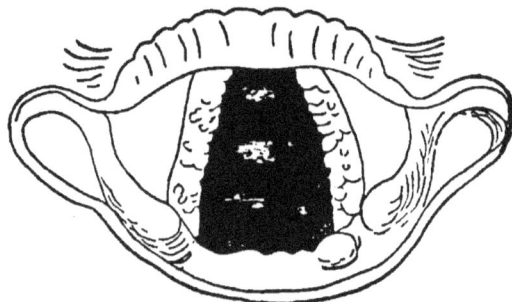

FIG. XIX.   TRACHOMA OF THE VOCAL BANDS.

The third picture (Fig. xx) shows the condition of the larynx of a young lady who had *lost* her *voice completely* for *years*, as a single look at the immense growth filling up the whole space will easily make you understand.

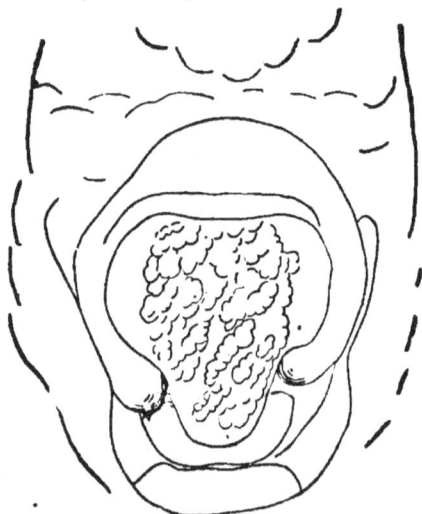

FIG. XX.   TUMOR GROWTH IN THE LARYNX.

None of these diseases could have been recognized, much less properly treated, without the aid of the seemingly insignificant, but really in its achievements grand and glorious apparatus, the little looking-glass. Death would inevitably have soon overtaken, for instance, the growth of this tumor, then only might it perhaps have been revealed. The laryngoscope enabled me to see it during life, to remove it and to cure the patient, who is probably in the audience this evening

---

Now we come to the question : "How does the throat enable us to talk and to sing?"

When any vocal sound is to be produced, no matter what is to be its loudness, pitch, quality or character in any respect, the vocal bands are made to approach each other ; the air, expelled from the lungs, strikes against them from below and drives them upward ; some air escapes, and the vocal bands, being very elastic, descend down to and below their previous level, then they go up again, and continue for a time to oscillate upward and downward. This up-and-down motion or vibration, occurring with appropriate frequency, is perceived as sound. Wherever there is such vibratory motion, there is sound, and wherever there is sound there is such vibratory motion. The statement that "sound" and "a particular mode of motion" are identical—one and the same thing—I shall try to make clear to you. Often the vibrations of a sounding body can be seen, as for instance those of a tensioned string like that of a guitar or violin : if I pluck the string of this monochord to one side, it rebounds and the vibrations constitute a sound. So far as *our ear* is concerned, the vibrations to be heard must occur at least sixteen times in a second, and not over 38,000 times.

The number of proofs and illustrations that can be given that sound is really nothing but to-and-fro motion within the limits indicated, is inexhaustible, but a few striking ones will suffice.

A tuning-fork can be made to give evidence on this point that is unmistakable and convincing. I have attached a little metallic pin to this tuning-fork (Fig. xxi) in such a way that

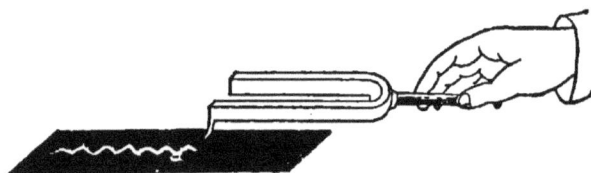

FIG. XXI.   AUTOGRAPH OF A SOUNDING TUNING-FORK.

if I draw it across this paper, it will make a straight line ; but the moment the tuning-fork is made to emit a sound, the line becomes a wavy one, thus testifying to the to-and-fro motion of the prong.   In order that this evidence may appeal to all of you, I will have this experiment repeated, and the line, made by a sounding tuning-fork on smoked glass, thrown upon the screen.

You shall now have a proof that the sound of the human voice is vibratory motion also.   I have here a glass tube closed at one end by a thin membrane carrying a little piece of looking-glass, the image of which can be thrown up on the screen. On speaking into the other end of the tube, the vibrations of

tne air communicated to the membrane will make the little
mirror move as you now see on the screen.

These experiments, which had been known a long time, sug-
gested, about twenty years ago, to the Englishman Scott, the
invention of an instrument which he called "phonograph" or
"sound writer." This instrument, a little modified and per-
fected by the accoustical instrument maker, Koenig, of Paris,
I have here to show you.

Here is a hollow truncated tone, at the smaller extremity of
which a membrane with a writing pen is attached in such a
way that it can be brought near to a revolving cylinder upon
which smoked paper is rolled. On speaking into the cone,
the membrane and pin are moved to make impressions upon
the smoked paper, just as in the two experiments seen on the
screen.

I have shown you that wherever we produce sound, we have
a vibratory or to-and-fro motion. Now I will prove to you
that wherever we produce a rocking or to-and-fro movement,
repeated with proper rapidity, we have sound.

FIG. XXII.—SINGING BAR OF COPPER PLACED ON BLOCK OF LEAD.

Here is a block of lead upon which I place this heated bar
of copper. At the point where the hot copper touches the
lead, this becomes heated and expands, tilting the bar away ;
the bar then touches the other side of the lead and is pushed

away, falling back upon the first side ; is tilted again, and so on. Thus a rocking or to-and-fro movement results which you hear as a singing sound.

I cannot conceive of a more convincing proof that sound is a mode of motion, except one *other* proof, which I will give you at the end of the lecture.

It is clear then that the up and down movement of the vocal bands produces a sound ; but that alone does not constitute the voice. The up-and-down movement of the vocal bands sets the column of air to dancing up and down ; the sound of these vibrations combines with that of the bands, and the sound of these combined vibrations, made more intense, mixed with various noises and changed in timbre in the passage of the air outward, through the mouth and nose, constitutes the voice. The sound of a column of air thrown into dancing or vibratory movement is easily illustrated in a glass tube or in any flue organ pipe.

As I have here two organ pipes tuned alike, I can demonstrate to you a very interesting acoustic phenomenon, although, it does not exactly belong to the subject. I refer to the production of musical beats. Both of these pipes give the sound of middle C, of 256 vibration, *i.e.*, when I throw the air in each into vibrations by means of the bellows, the air makes 256 vibratory movements in a second. By a little slide-valve arrangement here, I can change the rate of vibration, as I now do, of one of these pipes.

One of these pipes makes a few vibrations more than 256 in a second, and the difference between them you hear as beats. (Consider this experiment done in parenthesis, but it is really very instructive, and is the foundation for understanding the whole subject of musical harmony and dissonance).

If you will have the patience to hear how the throat and adjacent assistant organs of vóice and speech produce the variations of loudness of pitch, of timbre and of vowel and consonant sounds, I will tell you briefly :

1. Differences in loudness are mainly produced by differences in strength of the air current that is thrown against the vocal bands ; by this means the distance or space through which they swing up and down is made to become more or less extensive. You will readily understand how on this extent of swing (or amplitude of the vibration) intensity of sound depends. If I pluck this tensioned string of the monochord only a little out of its state of rest, the extent of its to-and-fro movement, or the amplitude of vibration, will be little, and therefore the sound not loud ; but if I increase the extent of swing, the sound will be louder.

Again, if in this bell, the sound of which is made by the particles of metal moving in and out, I tap the surface gently, the particles move only a small distance in and out, but if I strike it hard, the extent of their swing is larger, and consequently the sound is louder in this case than in the former.

2. Differences of pitch depend on differences of the number of vibrations that occur in a given space of time. The more numerous they are in a second, the higher is the pitch ; the fewer the vibrations that occur in a second, the lower or graver the sound is.

I hold a card against a revolving toothed wheel (see Fig. xxiii), and the faster it turns, *i.e.*, the more taps occur in a second, the higher is the sound. If I lessen the speed the pitch falls, and if it turns very slowly you can hear the separate taps.

The siren is an instrument with which we can count the taps or vibrations. Essentially, it consists of a perforated disk against one or more holes of which a current of air is blown (see Fig. xxiv). As an indicator shows the number of holes through which the air passes and the rate of revolution of the disk, you can easily determine the number of vibrations of any sound it makes.

FIG. XXIII.—CARD HELD AGAINST REVOLVING TOOTHED WHEEL.

In the case of the tuning-fork that wrote its autograph or phonograph for you, all that you would have had to do to find out the pitch,

FIG. XXIV. SIMPLEST FORM OF THE SIREN.

would have been to note the time it took to write the line and to count the number of angles or points in it. In the particular one I used, there were forty-eight points made in a quarter of a second ; therefore the sound of the tuning-fork is one of 192 vibrations in a second, or a tenor G.

In chords, pitch varies with the degree of tension and the length—the more tense the string is, the higher is the sound ; the less tense, the lower ; and the shorter the string is, the higher the pitch ; the longer, the lower.

In the vocal bands the variations of pitch are produced by changes of tension, of length, of breadth and shape, and of expiratory force. Some persons avail themselves of some of these co-acting factors to the exclusion of one or several of them ; some persons use different mechanisms under different circumstances ; some persons use prominently a factor which is not at all or but slightly used by others. The divergence of the views of several equally good observers as to how different variations of pitch are brought about, is doubtless due to this fact,—a fact entirely ignored by all scientific and lay writers on this subject.*

The lowest C on the piano is a sound of 32 or 33 vibrations' and the highest C one of over 4,000 vibrations in a second : a compass of 7 octaves, and most pianos have still a few notes more. The range of the voice differs considerably in different individuals ; commonly it is not less than one nor more than 2½ octaves ; in some great singers it extends to 3 and even 3½

---

* Since the public delivery of this lecture, a book has been published in which this fact has been recognized, viz., *Physiologie der Stimme und Sprache.* Von Dr. P. Grützner, HERMANN's *Hand-Buch der Physiologie*, vol. 1, part ii. Leipzig, 1879, p. 112.

octaves. The entire compass of the human voice exceeds 5 octaves, for there have been bassos as Grosser, Fischer, etc., who could sing the contra F of 40 vibrations with ease and power, and sopranos like Carlotta Patti and Christine Nilsson go up with ease to the high F of 1,400 vibrations, while the "Bastardella" and Mrs. Becker of St. Petersburgh could go up to and beyond the still higher C of 2,000 vibrations.

3. Timbre depends on the number and relative intensity of component tones mixed with the fundamental tone. This is the great discovery of Helmholtz who has invented a means of analyzing sounds by so-called resonators. (See Fig. xxv.) I

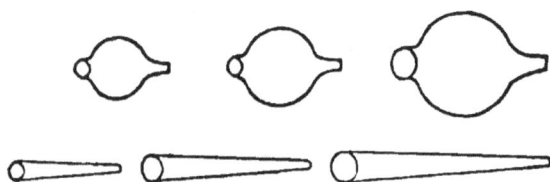

FIG. XXV. HELMHOLTZ'S RESONATORS.

have no time this evening to enter into details on this subject. It must suffice to say that by simply placing one of these resonators or analyzers into your ear, you can find out whether the tone to which it is attuned, is present or not in a mass of sounds ; and, that it has thus been discovered that every ordinary sound of nature is a composite one, consisting of a fundamental tone, and a varying number of other tones variously mixed with this fundamental tone. If Helmholtz had accomplished nothing in science except this invention and this discovery, his name would nevertheless forever shine in immortal glory !

Varieties of timbre of the human voice are brought about by

variations of the form of vibration of the vocal bands and by varying resonance, or co-sounding, of adjunct cavities.

4. If the sounding breath is not obstructed in its outward passage from the larynx, a *vowel* results. In this case, the cavities of the throat, mouth and nose, act simply as the resonance pipe of the organ.

When the nasal cavity takes a particular part in this resonance, the vowel is nazalized; otherwise the vowel- is pure. Variations in the relative size and form of the co-sounding cavities, especially of the mouth, produce changes in the *character* of the vowel sounds as *ah, a, e, o, oo*. I have placed those of the English language upon a large chart.

To form a *consonant* sound, the outward motion of the air is obstructed by a partial or complete contact of parts of the mouth and throat, the degree of obstruction varying from the slightest narrowing of the channel to an entire stoppage. There are some consonants which do not require the laryngeal voice at all, but are produced entirely by sonorous vibrations originating in the mouth and throat. We may therefore divide all the sounds of speech into two classes : (first) intoned, which originate in the specific vocal organ, the vocal bands, and include all the vowels and a majority of consonants ; and (second) unintoned, embracing the consonants formed by the mouth and throat alone, in the production of which the larynx plays the part of air-conveying tube simply. According to the parts of the mouth and throat, by the approach or contact of which the consonant sounds are formed, we may distinguish three classes of sounds, namely : first, *labials*, formed in the first articulating region—the lips : second, *dento-linguals*, formed in the second articulating region, namely, the teeth, tip of tongue and anterior part of hard palate ; and third, *palato-gutterals*, formed in the third articulating region, namely, the root of the tongue, posterior part of hard palate, soft palate and pharynx.

Analogous to the vowels, we have pure and nasalized *consonants*. In the *pure* consonants, as in pure vowels, the nasal valve is raised and closed. The nasalized consonant sounds have been called resonants. Here the nasal valve is lowered and open ; they are *m*, *n*, *ng*. In this diagram I show you a *schema* of the three regions of articulation and a view of the consonants formed in each. The mechanism of *m* and *p* is the most simple. The child, even the deaf from birth, learns to make these sounds most easily. You may observe in the chart that the main difference in the organs of speech between these two consists in the open or closed condition of the nasal valve. They are formed in the first region of articulation—the lips ; and the other organs are but little different from their natural position at rest. The lips need only to close and then to open to expel the air ; *p* will be heard when the nasal valve is closed, and *m*, when it is open.

The imitation of the words "papa" and "mama" in speaking dolls, is produced by a mechanism which I here show you.

Many more serious attempts have been made to produce artificially the sounds of human speech. The earliest was that of Kratzenstein just one hundred years ago. He was followed as to an apparatus producing vowel sounds by Willis, and more recently by Helmholtz.

More complete talking machines were constructed by

Kempelen and by Faber. Faber's talking machine was brought to this city about eight years ago, and for several years was exhibited in various parts of the country by a nephew of the original inventor, Joseph Faber, of Vienna. It imitates the human voice and speech organ in its construction. and being worked by means of a bellows and finger board, it can produce words and sentences in several languages with varying degrees of loudness, even to whispering, and with varying pitch. Still more recently a machine has been invented for *reproducing* instead of producing the human voice. This is the phonograph devised by Thomas A. Edison. Edison's phonograph differs from the phonographs I have described to you essentially in this, that instead of smoked paper or smoked glass, tin-foil is used for receiving the impressions made by the writing pin. The cylinder on which the tin-foil is placed, is grooved, and the metallic pin pressing against it, when moved by the membrane, makes permanent indentations in the tin-foil. *It occurred to Edison* [and it is this that marks the difference between men of genius and us common mortals, that all sorts of wonderful and unheard-of things, though they be really as simple as the indenting of Columbus's egg, occur to them and not to us], I say, it occurred to Edison that the process by which the indentations were made, might just as well be *reversed.* The motions of the air, called voice sounds, had made the membrane vibrate that carried the little pin that made the indentations, as the tin-foil was passed by. Now, to reverse the process, the indented tin-foil was passed by, and as the metallic pin maintained its contact with the surface of the tin-foil, it moved, just as it had done before. The membrane to which it was attached vibrated, and the motions thus communicated to the air, were heard as sounds.

This is a proof that sound is a mode of motion, of which 1 said that it is still more convincing than the singing copper bar which I showed you.   You see that Edison's machine is more than a phonograph, or sound writer; it is really a reproducer or regenerator of sound, for which a proper name would be "*Palingenophone*."   This is the name by which it should become known.   Hitherto it has simply been called a phonograph.

Through the kindness of Mr. Johnson, the secretary and manager of the company, that owns the patents and manufactures the phonograph, or *palingenophone*, I am enabled to introduce it to you and to let you hear it speak for itself.

The indentations made on the tin-foil are not easily effaced, and a speech once recorded can be repeated a large number of times.   The tin-foil may be put away, or sent to any part of the world or kept for any number of years; and thus the *words* and *voice* of a public speaker, a celebrated singer, or a personal friend or relative, alive to-day, may be heard by generations "yet unborn."

You may remember the beautiful rhetorical figure of "frozen music," and the story told by the Baron Munchhausen, of blowing into his trumpet and the notes freezing, so that in a few days afterward when a thaw came on, the music was heard.

All this and more, Edison's genius has realized.

COMPRESSED TABLETS

—OF—

CHLORATE OF POTASH and BORAX.

TWO AND A HALF GRAINS OF EACH; FREE FROM ANY ADDITION OR EXCIPIENT.

We ask the attention of Physicians to the above excellent combination, which will be found highly efficient in the relief of diphtheritic affections of the mouth and throat, and other morbid conditions of those parts, attended with disordered secretions. The depurative effects of these remedies are well known.

As the taste is not disagreeable, we have prepared them in the form of Compressed Tablets, thus giving the patient the benefit of their action, undiluted with Sugar, Gum or other vehicles, which would not only prevent their effects, but which sometimes themselves offend the stomach.

If allowed to dissolve in the mouth, the topical effect is much more efficient than a saturated solution, as while the solution is but temporary, the tablet really acts as a continuous gargle.

These Tablets have the great advantage over the gargles so commonly prescribed, that their ingredients are gradually dissolved in the saliva, and are thus constantly brought in contact with the affected parts. It must be evident that better results may be looked for from this, than from the momentary and occasional use of a gargle, which, moreover, is disagreeable to a great many persons, and to some impossible.

Children take the Tablets readily, as they have no unpleasant taste; while the convenience of carrying them in the pocket, commends them to travelers.

For Offensive Breath, they will be found equally efficient as our Chlorate of Potash Tablets, the addition of a mild alkali like Borax, increasing the activity of the Chlorate of Potash.

COMPRESSED TABLETS.

CHLORATE OF POTASH WITH MURIATE OF AMMONIA.

$3\frac{1}{2}$ GRS. CHLORATE OF POTASH, $1\frac{1}{2}$ GRS. MURIATE OF AMMONIA.

The advantages of the combination of these two efficient remedial agents, over either one administered alone, in certain conditions of the above ailments, will be readily appreciated by medical men.

The proportion of Muriate of Ammonia is so small, and being intimately mixed with the less soluble salt of Chlorate of Potash, the objectionable taste is not so apparent, and the medicinal effect really just as potent.

The claims advanced, when we first introduced Chlorate of Potash Tablets to Physicians, for the increased effectiveness of its administration, by allowing that salt to dissolve in contact with the inflamed surfaces, seemed almost an exaggeration; but since that time, the Tablets have been so largely used, Practitioners find the therapeutical results in the small doses given, more than borne out in the benefit to their patients, demonstrating really, that the topical effect and direct action upon inflamed surfaces, can be treated by this mode of prescribing, as efficiently as local applications on wounds upon the body, thus proving the advantages over constitutional treatment.

We wish to impress upon Physicians, the fact, that these Tablets are simply the pure salts, compressed without the addition of any excipient, sugar or gum, in this way avoiding any substance that would sheathe the surface and prevent the action of the salts.

They must not be confounded with the ordinary Lozenges, made by druggists, confectioners, &c.

DIRECTIONS.—Adults should take one every hour or two until relieved, allowing it to dissolve slowly in the mouth. Children, half of one as often.

In addition to the above, we manufacture Compressed Tablets of Chlorate of Potash alone, of five grains each, and Compressed Muriate of Ammonia, three grains each, thus enabling the practitioner to select the Tablet most strongly indicated.

JOHN WYETH & BROTHER,

Manufacturing Chemists, Philadelphia.

It will give us pleasure to furnish, on application, sufficient of these Tablets to test their merits by actual use.

www.ingramcontent.com/pod-product-compliance
Lightning Source LLC
Chambersburg PA
CBHW022011190326
41519CB00010B/1474